Mobile App Trends in 2015 For Mobile App Marketing

Vindimear D Heart

Mobile App Trends in 2015 :
For Mobile App Marketing

Vindimear D Heart

Mobile App Trends in 2015 : For Mobile App Marketing

Table of Contents

Introduction

As we move into 2015 with lots of wishes and expectations, one thing that also follows us is: people are increasingly moving from the disconnected world to the continuously connected one. People all around the globe are buying and adopting smart devices with a great pace, trying to change their ways of approach to life – changing the way we entertain, interact, educate and shop. According to a survey it is expected to continue with around 50 billion connected devices expected by 2020.

For the past 15 years, enterprises and consumers have been using mobile devices primarily to communicate and collaborate. Today's mobile devices are pocket-able supercomputers, and when they're combined with broadband networks and public cloud services, they become extremely powerful business tools.

Next year will be the first year that companies mobilize revenue-generating processes. Retail and Banking led this charge, moving large processes, like shopping and check deposits, to Smartphone and tablets. This move will continue to expand into other industries as the Smartphone and tablet replace the PC as a person's primary compute device.

It will also be the year the wearable becomes a mainstream consumer and enterprise device. "Data Snacking" is the ability for a person to consume small, frequent amounts of data on a smart watch. This phenomenon will continue to grow as watchmakers cater to smaller wrists and high fashion demands. These devices will start to be used as part of multi-factor authentication schemes as well.

The mobile app development industry is thriving and continuing to evolve year after year. In 2014, we saw mobile app market maturing from Smartphones and tablets to wearable devices and Internet of Things. There was also an increased focus on app analytics and mobile app marketing. Now that 2015 is around us, we decided to list down the top mobile development trends to expect in the New Year.

Here are some of the predicted mobile trends for 2015:

1. Mobile Apps associated with Wearable Gadgets

The mobile market will rise with upcoming brands of mobile apps which will be linked to Wearable Gadgets. The most common example is of Google Glass which has been a topic of interest for tech savvy people. Another example that comes to mind is of Apple Watch which again is another area where most of the technical minds are focused. With these 2 devices in the market hitting the charts, the mobile apps associated with them will also find new highs.

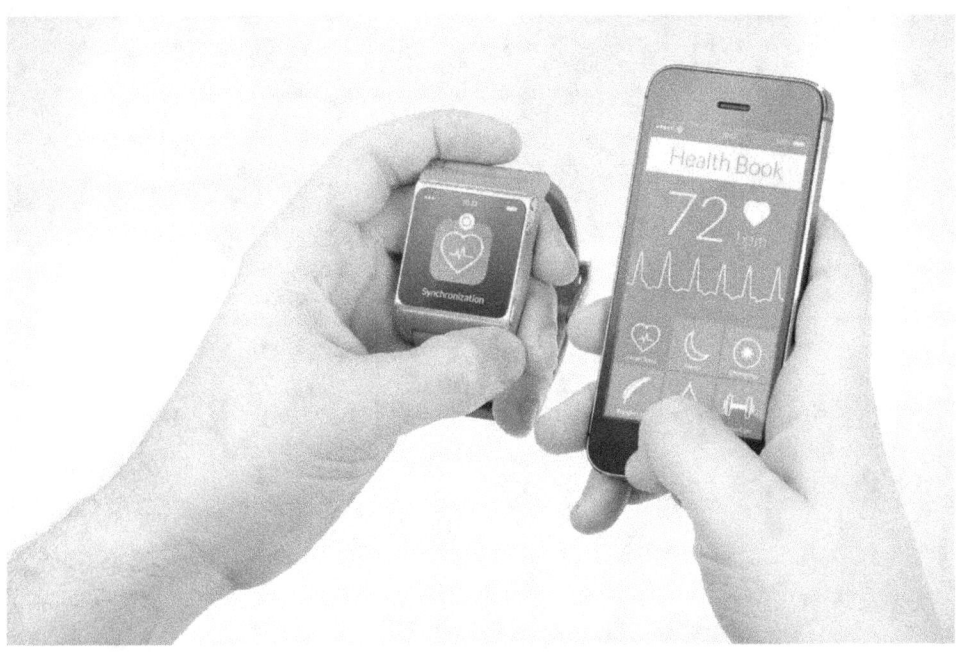

2.   Social Networking Apps

A recent study states that one quarter of the world's population uses social media. This means that 1,730,000,000 people are posting, pinning, tweeting, snap-chatting and instagramming. Every 60 seconds 4.7 million posts are uploaded to Tumblr; 277,000 snaps are shared on Snapchat; and more than five million videos are viewed on YouTube.

Social websites and apps being among the most used features on mobile, 2015 will inevitably see optimized social media apps with integration of new features.

3. Apps Managing Your Home Appliances

The competition for a connected home with a unified access to all the appliances will rise. The Mobile Apps will control and configure home appliances. The unveiling of the Apple's HomeKit product seven months ago led some enthusiasts to believe that by the end of 2014 they'd be remotely controlling all manner of electrical items – including lights, locks, thermostats and plugs – via their iDevices or Apple TV. But Apple seems to be moving with a snail's pace and according to me it will come into market in spring this year.

4. Identification Apps (Touch IDs, etc)

As Apple Inc. have designed and released Touch IDs in iPhones starting from iPhone 5s and iPads, 2015 will also witness amazing Identification technologies and mobile apps associated with them. Apple says Touch ID is heavily integrated into iOS devices, allowing users to unlock their device, as well as make purchases in the various Apple digital media stores (iTunes Store, the App Store,iBookstore), and to authenticate Apple Pay online or in apps. Mobile Apps using these technologies will also find good grounds in market. These Apps will make the mobiles more secure and personal.

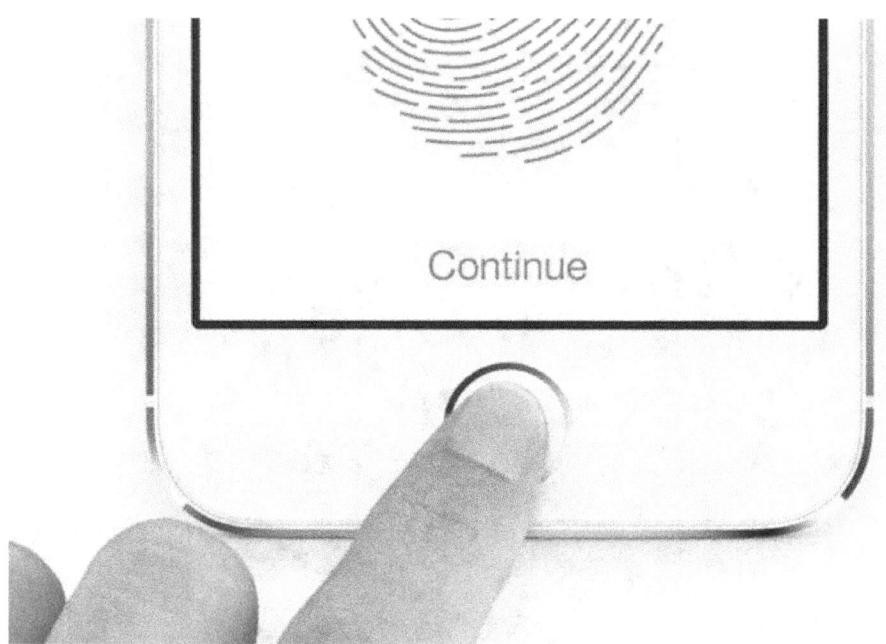

5.   Vehicles - Mobiles Integration Apps

As the mobile apps will be linked to Wearable Gadgets and Home Appliances, they will also find their way into the Vehicles. "GPS systems installed in cars will be replaced by in-vehicle Wi-Fi, which will keep the location and vehicle information updated" said Stu Lipoff, IEEE Fellow and Engineering Consultant. These apps will provide a better and enhanced form of communication with your vehicles so that you can know better about your rides and can take good care of them.

6.   Device agnosticism will increase

"Forrester Research" found that 90% of users who own multiple devices start a task on one device and finish it on another. Without a doubt, the one trend we see informing almost all other trends is device agnosticism. Apps are increasingly becoming experiences that live across multiple platforms — from wearable to phones, tablets, and web applications." As this trend proceeds in 2015, offerings that can seamlessly transfer between these states as you move from one device to the next will have a huge advantage.

7.   Health and Nutrition Monitoring Apps

Your mobile and wearable devices will generate real-time data regarding your individual body — tracking blood glucose levels following scheduled meals, sleep quality, carbon dioxide levels in your muscles, blood pressure etc. Not to mention smart armbands connected to mobile apps, for workout-related notifications, and smart shirts that can notify you about stress levels or an elevated heart rate.

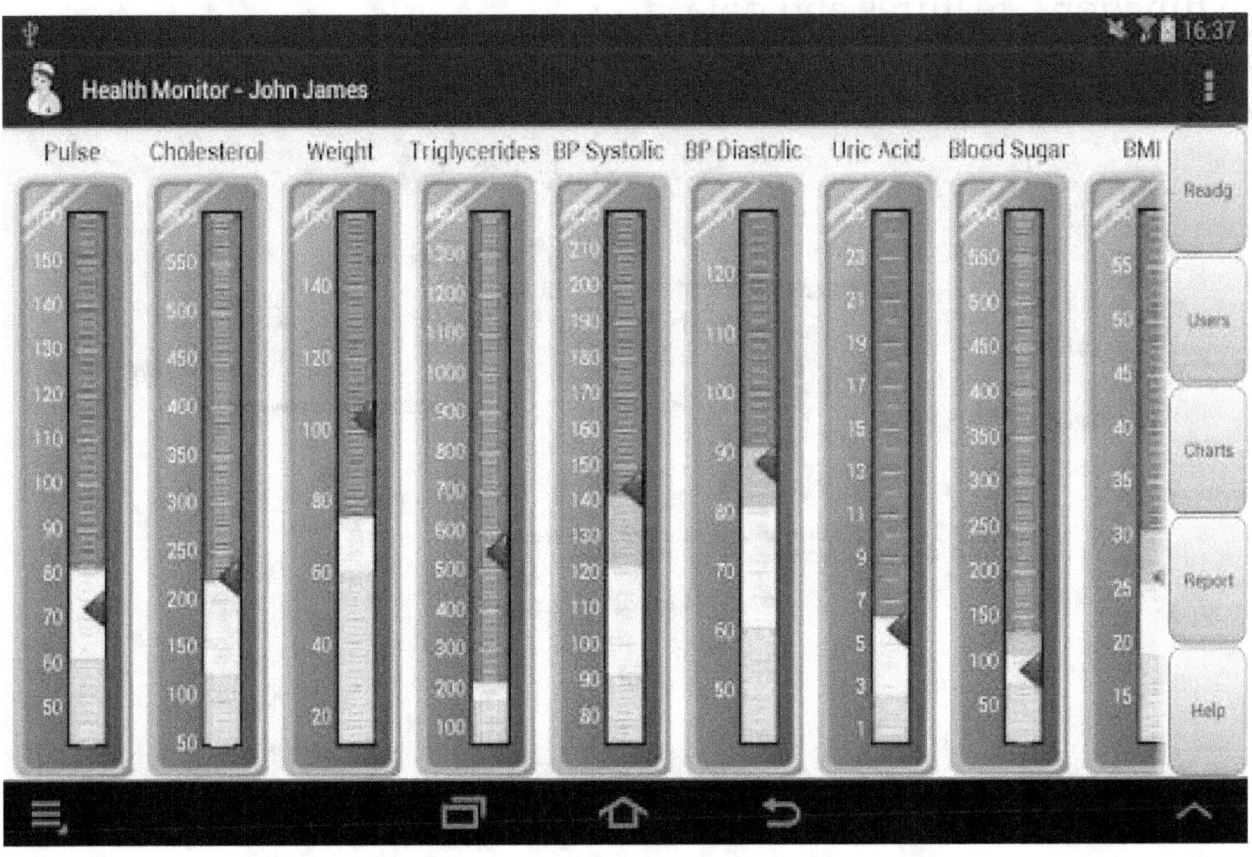

8.   Cloud Driven Apps

The cloud technology will play a key role in mobile app development in 2015. With the rise in the usage of multiple mobile devices and wearable tech as mentioned under the device-agnosticism heading, app developers will have to focus on the ability to integrate and sync their apps on multiple devices. The cloud approach will enable developers to build their applications that can be accessed on multiple devices with same functions, features and data.

9. App Security

There were numerous reports related to hacking and information leaks in 2014 and you all know about the celebrities' leaked pictures. As per Gartner's prediction, 75% of mobile applications will fail basic security tests in 2015. The hackers will continue to exploit security gaps in mobile applications to crack sensitive information. Security remains a big challenge on mobile devices. Key point, mobile app security is something developers need to seriously act upon in 2015.

10. Internet of Things (IoT)

Internet of Things is increasing with more and more people connecting via multiple devices. Apps will further mature as they get integrated on multiple connected devices. It will inspire app developers to focus more on user experience on various devices. With IoT, the customer engagement management platforms based on Wi-Fi or Beacon technology will also get a breakthrough to help businesses reach consumers via apps at the right time and right place.

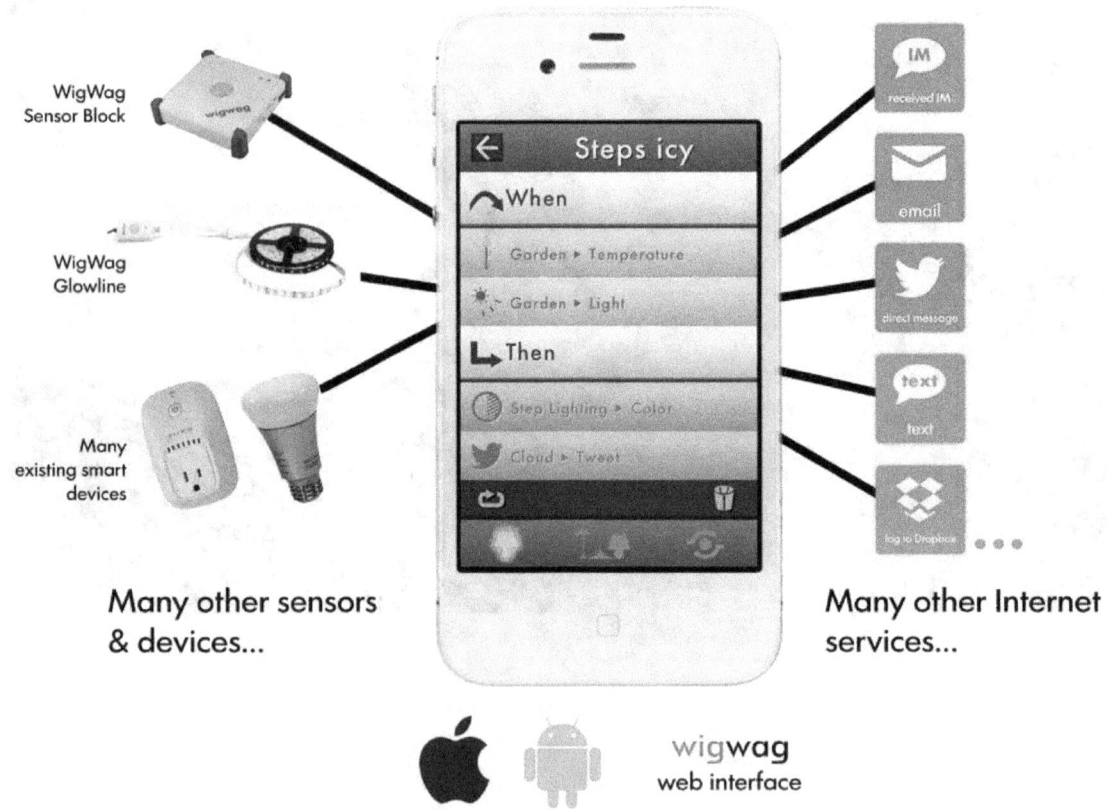

11. In- App Advertising and Purchasing

According to a new study from Juniper Research, a research firm, in-app advertising spend across all mobile devices will reach $7.1 billion by 2015 and that's huge. Mobile Ad networks will go beyond banners and experiment with various new ad formats. Mobile app advertising and in-app purchases have already revolutionized the money making techniques and the developers will continue to focus on them as it has become a primary source of monetization and a key to success as more and more app developers are shifting away from the paid download model.

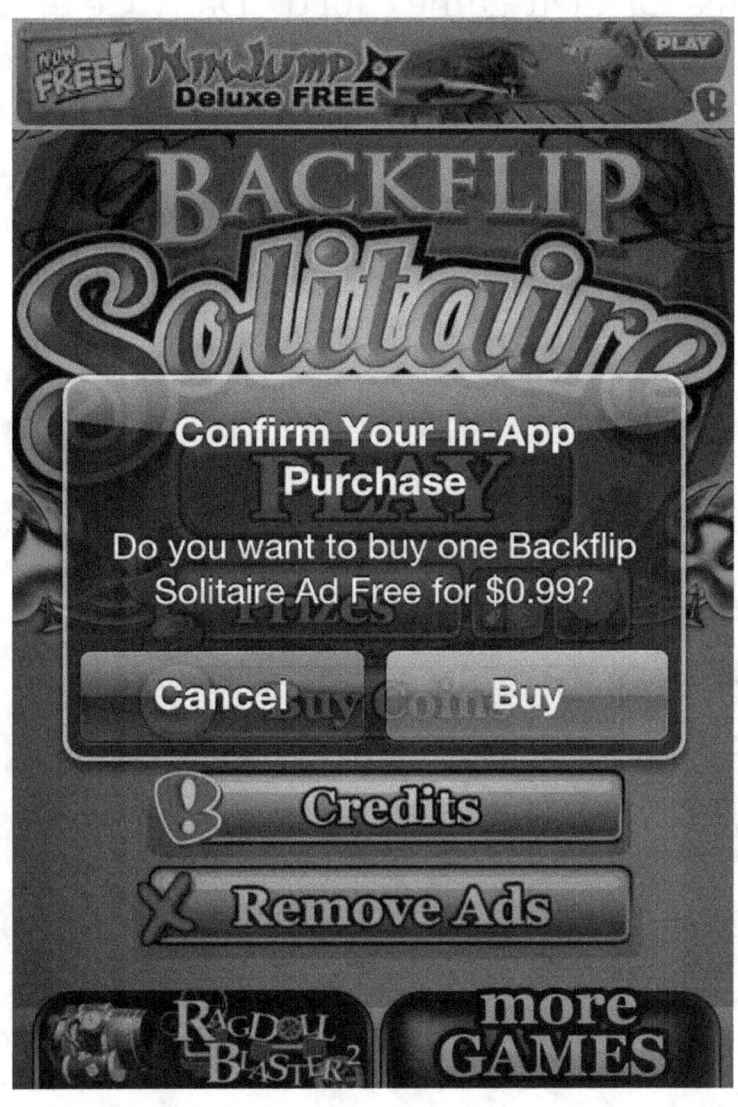

12.   HTML5 Apps

HTML 5 and its development tools have been gaining popularity in 2014. As this technology mature, more and more businesses will adopt it to create hybrid mobile apps that work on multiple platforms with ease. In 2015, the focus on native apps is expected to decrease with the rise of hybrid apps based on HTML 5.

Another interesting trend regarding HTML5 applications is its use for marketing other Apps. PuzzleSocial, a game development studio based in NYC, unveiled a case study about leveraging mobile web to drive high-quality installs for their game called "Daily Celebrity Crossword", which is available for iOS, Android and Kindle Fire. They built a lite version of the game on HTML5 and distributed it via mobile web. In 4 months, this resulted into 320,000 unique game plays and, which is more important, 10% of those players proceeded to download the native mobile app.

13.   Mobile Gaming Apps

In 2014, the mobile games have gradually started to move from single player to multiplayer. With the rise of multiplayer games the social interaction and social media integration within the mobile games will become more important than ever. In 2015, the mobile games are also expected to be more cloud driven due to the demand of frequent and regular updates in the games. This will require ongoing management by the game developers.

14.   M-commerce, Banking and Mobile Payment Apps

According to a research, almost 19% of ecommerce sales in 2014 were made on a Smartphone or tablet. Many believe this trend will continue over the next 4 years as more and more consumers adapt to m-commerce. Using a mobile phone to pay or purchase instead of debit or credit cards will also become more common in 2015 with Apple Pay and Google Wallet. This will allow developers to build mobile apps that can process transactions without the need of physical debit/credit cards or cash.

Whether it is a one-click checkout on any e-commerce site online or the ability to click a button on one's phone just before walking out of a physical store, the brands that fully embrace the utility of mobile payments will be big winners in the end. The biggest potential value here is the ability to stitch together online and offline profiles.

15. Travel Booking Apps

"Companies such as TripAdvisor, Hipmunk, Skyscanner, Trivago and Dohop are going to make it increasingly easier next year to book flights and hotels right within their apps instead of sending consumers off to airline, hotel, or online travel-agency websites to complete their bookings," said Dennis Schaal, news editor at Skift. "Another thing is that Expedia, Hipmunk, and others are increasingly making it easier to start your trip research on a Smartphone, continue it on a laptop, and then pick it right up again on a tablet or Smartphone — right where you left off."

Social media, content marketing, influencer relationships, and creative events have proven their influence and 2015 will be the year that companies across the travel spectrum commit to a specific lifestyle message with the digital and real-life assets to back it. As companies tap into their underlying story and create long- lasting connections with customers, they'll be able to better communicate via mobile apps and social feeds to become a part of aspirational travel from the start.

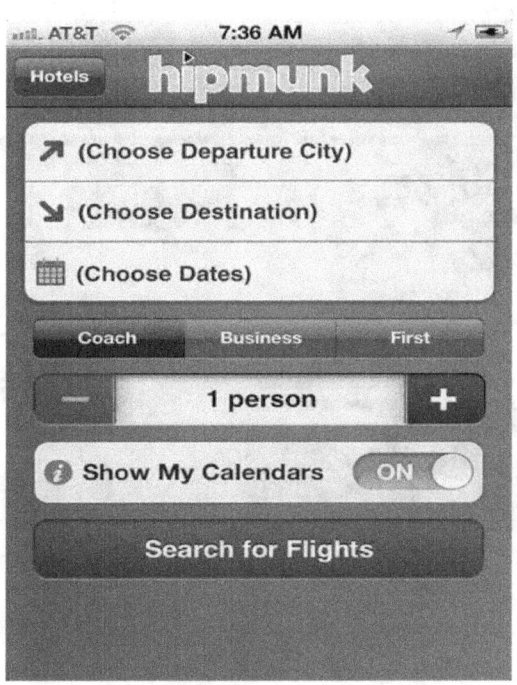

16. User Experience First

User experience will lead technology in 2015. With the increasing number of Smartphone, tablets and wearable devices, the app user experience will be more critical than ever. We will see more advanced and easier to use interfaces, more powerful support of swipe from bezel operations and more support and usage of voice recognition in most applications. With more focus on in-app advertising and purchasing mobile app user experience will be more challenging.

Analytics data will play a major role in helping app developers create better user experiences. These new user interfaces will play a critical role in the publicity and purchase of the mobile apps. Location-based services will continue to thrive, likely with more social and m-commerce discovery. And it will become easier for enterprises to build mobile applications, as a number of codes free; end user friendly mobile app creation platforms will hit the mainstream.

17. Big Data and App Analytics

As Internet of Things and Wearable market grows, the need of app performance tracking and analysis will grow among the decision makers in various enterprises. There will be more focus on big data and analytics in 2015. Mobile app developers will continue to focus on adding new data collection methods in their apps to get more insights and actionable items to meet their client's expectations and make more engaging and successful apps.

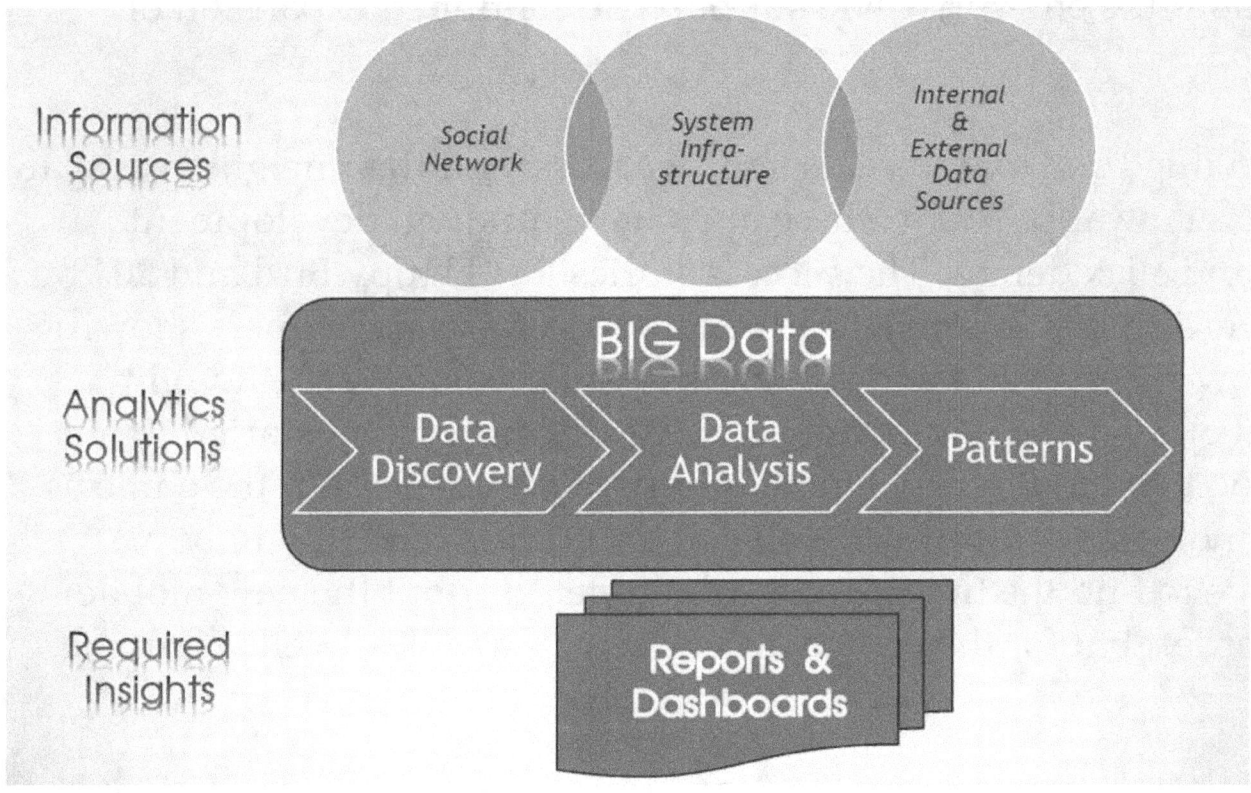

Conclusion

Clearly, there are other major trends happening in mobile in 2015, like an increased focus on mobile for internal collaboration. Mobile will also play an increasingly important role in the evolution and user behavior of social networks both new and existing. We will also find out what impact more wearable devices like the Apple watch and Samsung's Gear series will have on consumer behaviors and purchase patterns. Increasing app usage, greater end-user sophistication coupled with improved user experience, and strong initiatives to go mobile throughout the enterprise will support the continued expansion of mobile.

Just thinking back to two years ago, everyone and their neighbor had an idea for a new app. Today, these apps have funding, development teams, and slick demos. The success stories like Flappy Bird and 2048 alone were an inspiration to this generation of app developers showing them how far an original idea can take you. If you haven't thought much about your mobile strategy, now is the time to get started. It's never too late but it will get increasingly harder to get into the game as the technologies continue to evolve at lightning speed. It will be a challenge to find the time, focus and budget to adopt the new mobile world order, but for those who do, the rewards will be well worth the effort.

THANK YOU

Author

My name is Theeradech Thapanaphong and I have experience in many fields as Medical field, Astrology Field, Career Field, Outsource, and Consultant. I shared all my skill and experience to this book and we want to give my knowledge to all the audiences.